The Social Media Starter Kit

Kerry Rego

Kerry Rego Consulting Books
Santa Rosa, CA

Kerry Rego Consulting Books
1011 2nd Street, Suite 100
Santa Rosa, CA 95404
www.kerryregoconsulting.com

Book Layout ©2013 BookDesignTemplates.com

Ordering Information:
Special discounts are available on quantity purchases by corporations, associations, and others. For details, contact the Kerry Rego Consulting Books at the address above or book@kerryregoconsulting.com

The Social Media Starter Kit / Kerry Rego. —1st ed.
ISBN 978-0-9906183-0-0

Table of Contents

Introduction

Content Creation

- Types of content
- Build a house of social media
- Editorial calendar
- Balancing your content
- Where to get ideas for content
- Follow the rock star
- Team communication
- Content receptacle
- Public relations assets
- Establish vendor partners
- Helpful tools for content creation
- Content use cycle
- Search engine optimization (SEO) and being found online
- Images and copyright

Timing, Scheduling, and Automation

- How frequently to post
- Scheduling tools
- Benefits and dangers of scheduling
- Automating posts
- Labor hours

Measuring Performance

- ROI and how to tell if it's working
- How to build a metric report
- Channel analytics
- Reporting and review
- Record keeping
- Dark social and the immeasurable

Policies, Crisis, and Reputation Management

- The importance of having a social media policy
- Listening
- How to deal with negative online feedback
- Taking responsibility
- Common user question
- Why deleting negative posts is a bad idea
- User story
- Creating a crisis plan
- How to get manage your online reputation

How to identify the right team member for social media management

- Skills for the job
- Training
- Outsourcing the job

Best practices

Resources

- Templates
- Statistics and information
- Marketing
- Media
- Online training

Bibliography

Dedication

Everything I do, I do from my heart.

Joy & Daniel,
Thank you for all your love.
I couldn't do it without you.

CHAPTER 1

Introduction

I started my organizational consulting business in 2006. I helped my clients become more efficient with their work processes and technology use. By 2009, I was working exclusively on social media and technology needs for my clients due to the drastic growth in the tech sector with mobile computing, social media, and cloud software. No one could have predicted that social media would arise and become so dominant but I like to say I had the right skill set and passion at just the right time.

As the years have progressed, the industry has shifted significantly, my clients' needs have changed and, to be honest, I've gotten better at my job. I've collected a tremendous amount of information that I share with my clients as we work on their projects. (I call them "cheat sheets.") It got to the point where I simply needed to bind them.

As I wrote this book, I imagined that I was sitting next to someone just like you, guiding you through all the steps on how to create your own social media strategy. With this book, I've attempted to duplicate the process that I go through with my clients on a daily basis. I have learned that as my clients' companies grew, I wouldn't be able to train everyone they hired. If I were to leave them with a guide to walk them through it, what would it look like?

I could see it so clearly but I needed confirmation that I'd structured it right. When I taught a one-day class named "Social Media Starter Kit" at Santa Rosa Junior College in the Fall of 2013, I was able to see the natural flow of the entire process in one sitting. Within 24 hours of teaching that class for the first time (I teach it now in multiple places), I wrote the table of contents for this book and was ready to get started.

Just weeks later, I was contracted to write a social media manual. I stopped to focus on that project for 6 months and it helped me see the material from a yet another perspective. This book is the original Starter Kit idea with the feel of a manual. My goal is that with this book you will be able to sit down and create your own effective and sustainable social media plan.

How to Use The Social Media Starter Kit

I designed this book to guide you and your team through the process of setting up a successful social media campaign and to be a useful resource for you. You may be a sole proprietor, small business owner, marketer, or large organization. Whether you are from a for-profit, non-profit, educational body, government, or otherwise, I believe anyone can use this information. Please bear with me that I'm not able to use every form of pronoun in each instance. When you read "your organization", it's directed at a single person entity or 1000+. I use this general address because I feel it works for all. In the case of my work, I consult with the full gamut I've just stated and what I share is a result of my experience with my clients.

This book includes video content. When you see a QR code (that square bar code below, I explain more about Quick Response codes in Chapter 3: Tools, Accounts, and Digital Assets) in the printed version, scan it to access videos so that you can hear from me personally. If you haven't scanned a QR code before, simply download a free QR code reader app from your mobile app store. Search for "QR code reader" and you will get several options, it doesn't matter which you choose.

Disclosure: I am not sponsored by any of the companies you see mentioned in this book. I'm not receiving compensation for recommending any of the products you read about here.[1]

If you are reading a digital version of this book, your copy will have a Starter Kit image which when clicked will take you to the videos. For both versions you will need an internet connection to watch the videos as they are hosted on the web.

I have provided links to templates listed in the Resources section so that you can build your strategy as you work your way through the book. I've also created a bound workbook with the same information should you prefer that format. These are great for employees and teams or as an easy to access supplement to this written work. You can get a copy of it on my website http://kerryregoconsulting.com/store/.

Occasionally, I've written **Stop and Write** to remind you that you aren't just reading this book, you are using it to guide you in your social media strategy. Remember, you **can** hire me virtually or in-person to work with you; this book is a perfect companion to all that we'll do together. Many of my clients will be happy to hear I've written this so that they have a little of our sessions in the bookshelf whenever they need it.

I have to clearly state that social media is an industry that moves rapidly. Tools are phased out or bought by other companies while demographic information and best practices are subject to change. The bottom line is, **things change** and you will need to accept that fact. I have tried to provide you as many resources as possible to adapt and update your strategy. Be prepared.

Best Practices can be seen in many places throughout the book to highlight important details.

CHAPTER 2

Build a Strategy

M ake sure you are in the right place. Just because everyone is using social media, doesn't mean you have to. *What?* You read that right. I am a social media consultant who's telling you social media may not be the answer to your problems. Ask yourself this one question:

"What tools does my intended audience prefer?"

When I asked this question of a client once, I quite accidentally learned one of the most important lessons in communication. I was having trouble reaching her on the phone so the next time we met for an appointment, I asked her, "How do you prefer I communicate with you?" She answered, "By email. I never answer my phone." I was dumbfounded. I thought back and realized every time that I was successful in talking with her was via email. This was the watershed moment when I knew there was no assuming – everyone has a preference and I have to ask.

It's very possible that your target audience or customers would rather you call them on the phone or send them physical mail. You know your customers really well, if you doubt how well you know what they like, ask them. You can also research their online habits at Pew Research Center's Internet and American Life Project (otherwise known as Pew). The Internet and American Life Project is a division of Pew Research Center, "a nonpartisan fact tank that informs the public about the issues, attitudes, and trends shaping America and the World". It is considered the data source for people and their behaviors on the web.

I would hate for you to be spending your time and efforts on the wrong methods of reaching current and potential customers or those that would be interested in your services. Make sure they use social media before investing time in building and deploying these tools.

Social Media is Not Magic.

Social media consists of internet-based applications, tools, and platforms that turn broadcast monologues into dialogues. There are many definitions but the main parts are they must be on the internet, allow conversation, and be shareable.

Social media isn't magic. It's best use is not selling your products. The word "sales" is not in the definition. It can definitely help you increase your sales but it doesn't replace all forms of marketing, sales, or promotional endeavors. It's a supplement to what you are already doing and may upend your marketing budget when you find it is particularly useful or fruitful for your purposes. It is not completely free and will require labor hours. It will not fix a broken business model. If you have big problems in-house, social media will not perform miracles for you (well, it might but don't bank on it).

Part of my job is to manage your expectations around what is possible. Social media works best when communicating with the public and driving traffic to your conversion sites (where you make money or provide services) and we'll talk more about types of content you can create in Chapter 4: Content Creation.

User Story

I had a consultation years ago with an attorney in my hometown. He was very interested in creating social media plans for several businesses and excited about the possibilities. I clearly explained to him the services I offer: creating strategies, working with teams to implement, and training staff. He asked me if I managed accounts and I regretfully said no, that I was a trainer not a maintainer. He thought for a minute then asked, "But can't I just pay you to do it?"

It felt as if time slowed down. This is the moment I first recognized that some people prefer to purchase their branding. They believe if you pay top dollar social media is a bottled product you can have delivered to your doorstep. I prefer to think of social media like a woman: she's ever changing, the second you think you've figured her out, she changes her mind; you can't buy her love; talking is a strong trait; it's complicated; and there's no easy solution.

Social media isn't a product you can purchase.

The Importance of a Social Media Strategy.

The social media strategy is your road map for success (see Resources for format) and without it, you will flounder.

When I first got my driver's license, I'd jump in my car and drive aimlessly for hours, listening to music with the wind rustling my hair. I doubt your budget has money to spare for aimless wandering on the web or extra manpower for fooling around. You may have already experienced this but this is exactly what happens when you don't have a strategy. Now when I get in my car, I have an end destination in mind and a plan as to how I'm going to get there. Social media is no different. Get a plan and you will be productive instead of wasteful.

There are nine basic parts but may include more as needed:

- Profit centers, services offerings, value propositions
- Goals
- Target audience
- Messaging
- Content planning
- Tools
- Measurement
- Policies and crisis management
- Staff responsibility
- Reporting and adjustment

First Things First

Your organization makes money, offers services, or has a value proposition. You might have a combination of the three. Write down your answers to the questions that are applicable to you:

- For-profits, where does your money come from? How do you make money? You have one or possibly multiple sources of income. If you only sell one product, there are most likely different uses for it.
- Non-profits, what services are you offering? These may be programs, services, educational offerings, etc.
- What is your value proposition? What value are you providing the world? Why should they come to you?

Stop and Write. Now that you've written that down, you can move on to setting goals or why you are using social media.

Setting SMART Goals

There is a mnemonic acronym that will help you set your goals. The term "SMART" is regularly attributed to Peter Drucker when he used it in his management by objectives concept though it's first known use is by George T. Doran in *Management Review* released in 1981.[2]

When setting a goal for your social media endeavors, using this acronym will help you to set smart goals (groan, sorry) rather than ones that are based on ego. There are ways to measure activity that are known as "vanity metrics" just for this reason. So I suggest that when you're setting your goal(s), ask yourself if it/they fit(s) all these requirements:

- S – specific
- M – measurable
- A – assignable (or attainable)
- R – realistic (or relevant)
- T – time specific

Clearly state your purpose for using social media. Goals are expected outcomes of activities that are measurable and have a timeframe.

SMART Examples:

- Increase visits to website shopping cart by 20% in the fourth quarter of 2014.
- Increase email signups by 200 people by December 31, 2014.

- Achieve 100 mentions of brand promo code in October 2014.

Not so SMART examples:

- Increase sales
- Get 1000 fans on Facebook
- Get more "likes"

The second set is typically how goals are named, if any are stated at all. Let me show why they don't work. When an objective is vague such as "increase sales", there is no benchmark and there's nothing to measure to gauge success. "Get 1000 followers on Facebook" is measurable but it doesn't support your bottom line and it doesn't really mean anything. "Get more likes" is vague, immeasurable, you'll never be happy with what you've got, it doesn't support your bottom line, and there's no end in sight.

I often ask this question, "What would you like social media to do for you?" This can be helpful in getting to your true desires. Personally, I'd like it to wash my car, do my taxes, and my laundry but that's not realistic. Be realistic in your goal setting and the results come with planning and discipline.

Determining Your Target Audience

You know who buys your products or services. You know who you **want** to buy your products or services. List them out one by one and delve into who they are, how they feel, what's important to them, whatever information is useful in understanding their needs. Include information such as age, life stage (college-bound, new family, "empty nester", retired), gender, ethnicity, socioeconomic status, education level, cultural nuances, needs, location etc. Give them a name so it's easier on you to remind yourself whom you are talking to online. It may look like this:

Target 1: Moms

18-35, working moms, stressed for time, probably spend a lot of time commuting, low on sleep

Target 2: Office Managers

30-70, male and female, trapped in an office all day, underappreciated, stuck between the corps and the higher-ups

No, your audience isn't really everyone (I hear this almost daily). Who currently buys your products/uses your services? Who are you targeting? Who do you **real-**

ly want? You are going to have categories you can name and list. Narrow down "Everyone."

Stop and Write.

Messaging

This is the time to establish what message you want to get across to the public. What do you want them to know, believe, think, feel, or experience? What words do you want them to use when discussing your brand? It's up to you to clarify and write it down. We delve into this subject in Chapter 4: Content Creation.

The concept of keywords[3] is crucial for you to understand as it's how you will be discovered on the internet. Keywords are the words that people type or say to look for information. Keywords are read by search engines and categorize your content in order to serve it those that are looking for what you have to offer. Your website and all your social content should contain keywords that describe who you are, what you do, and why you do what you do. Think about the words the layman or the customer uses to describe you. You use technical and industry terms, they may not. When I noticed how people described me, it changed the way I did as a result.

As an example, here are some keywords that describe me, my services, and my content (whether or not I use them on a daily basis): social media trainer, social media marketing, social networking, educator, teacher, writer, author, columnist, blogger, speaker, instructor, consultant, Facebook trainer, and LinkedIn specialist, to name a few.

If necessary, put this on a sticky note or a printout, as well as on your Master Social Media Document to remind you and everyone else on your team to be using these words on a regular basis on your website, in your printed collateral, and in conversations on and offline.

In addition to having keywords, you will need to differentiate your message using your target audience list. Determine how your message and communication style will differ depending on whom you are attempting to reach. Last year I did an exercise in the social media course I co-teach at Sonoma State University and used a local dairy producer as the mock company for whom we were building a strategy. The messaging of "milk does a body good" for kids was designed to resonate with moms but we had an entirely different message for those that had gotten away from dairy and we wanted to woo back. We designed another for lactose intolerant customers as well as those looking for organic food options. One company, four audiences, four different messages.

Here's another tool for you to try. I have a process I call "The 5 W's." This is a guide I use personally and with my clients to help them to determine what their messaging should be:

WHO are you trying to reach? WHO are you to them? WHO are you?

WHAT are you trying to accomplish? WHAT do you want them to know?

WHEN is it important that they hear from you? WHEN do they prefer?

WHERE are they on the internet? WHERE are they in the world?

WHY is your account or message important to them? WHY should they care?

Planning Your Content

Content planning is an ongoing process. Much of the content and inspirations will be stored digitally so that you can easily move it around and share with team members. It will include: an editorial calendar, content ideas, calendar of events, partner programs for cross promotion, a traditional (print, radio, tv, events) marketing editorial calendar, and ad buying information (see Resources for several of these templates).

Worried about what you'll be posting? Don't. We'll talk more about this in Chapter 4: Content Creation.

Tools

This is your master list of accounts and digital assets. For each property you manage, this list will include: name of asset or tool, website address, log in information, whom has access to account, and current status (see Resources for Digital Assets template). I devote all of Chapter 3 to this subject.

Measurement

It is imperative to have metrics to show whether or not you are achieving your SMART goals. You will be collecting data for both digital and physical reports. Chapter 7 is where you will learn much more about this subject (see Resources for metric report templates).

Staff Responsibility

This is where you determine what team member will be responsible for which task. Seems simple but it's often overlooked. Delegation is key. This information can also be included in the Digital Assets List under Access.

Policies & Crisis Plans

You may not need an entire policy for social media use, it may be incorporated into your current business or employee manual but I provide resources for creating your own. When disaster strikes, man-made or natural, you will have a plan. Both Chapter 8 and Resources are places to get this information.

Best Practice: Make a plan. Clarify: profit/service centers, goals, target audiences, and keywords. Employ the WHO WHAT WHEN WHERE WHY method to encapsulate your strategy.

CHAPTER 3

Tools, Accounts, and Digital Assets

B efore you ask me, "Kerry, which tools should I be using?" I'm going to turn it around and ask you what your goals are for social media. The reason I know you want to ask me that question is because everyone does. Of course, my clients ask this question but I get it at cocktail parties, in board meetings, and in all types of casual conversations.

What problem do you need solved? Now find the tool to provide the solution you need.

What follows is a general list of popular categories and the tools that belong to them. Some have multiple functions or features and could be placed in more than one category. See Resources for additional websites where you can get detailed demographic information.

Email marketing*

- Example services: MailChimp, Constant Contact
- Features: ability to email to customizable quantities of recipients, emails have exemplary design features, embedded media, links to company website, social media sharing options, scheduling, segmenting audiences, unsubscribe management, performance analytics, in compliance with CAN SPAM Act of 2013.[4] (Please see Resources to read more about this very important ruling that if you violate you could owe the government a lot of money!)
- Demographics: 85% of US internet users access email every day (Reuters/Huffington Post)
- Importance: High, based on usage numbers
- Usage note: Highest point of conversion, still the most common form of communication, open rate is best on Mondays (worst is Tuesday), best times of day are 5-7am

Some argue that this tool doesn't adhere to the three basic features of social media. It drives conversion and is highly social. In order to achieve your SMART social media goals, I believe it is an imperative part of your strategy.

Blogs

- Example services: Wordpress, Blogger, Typepad, Tumblr
- Features: self-managed sites that allow for content publishing including text, audio, video, links, public commenting, social sharing, and the ability to link to other pages within blog or main website
- Demographics: average blog reader is 41, 45% female and 55% male, (Pingdom). Tumblr is unique in this group as 66% of users are under 35 (Search Engine Journal)
- Importance: High. Search engines trust and rank blogs sites favorably in search results. Blogs are a major source of traffic to websites.
- Usage note: Though subscription rates are down, blogs are excellent opportunities for long form message delivery, can be used to answer client questions, and it's the most common way that new visitors will find a website.

Microblogs

- Example services: Twitter, Plurk
- Features: shares same functions as blogs but are constricted by file and physical size
- Demographics: varies by service
  - o Twitter – 18% of online adults are users, 46% of users logging on daily, Black Non-Hispanic users and 18-29 are the dominant ethnic group at 29%, while White Non-Hispanic and Hispanic tie for second at 16% (Pew Research Center)
  - o Plurk – doesn't have enough of a U.S. presence to be considered
- Importance: Medium level of importance. The speed of use required by Twitter can be an obstacle for account managers to maintain.
- Usage Note: This is the toughest category of tool to maintain. The speed that content is distributed on these channels is often overwhelming for those responsible for creating content. Scheduling and time management are extremely important for this category. Plurk was created in Canada and has relocated to Asia because that's where the majority of their traffic originates and isn't considered useful to U.S. audiences.

Photo

- Example services: Flickr, Instagram, Pinterest
- Features: allows for photo sharing on the web, photo storage, search engine optimization, photo editing
  - Demographics: varies by service (Pew Research Center & B2B Infographics)
  - Flickr - 18-34 & 55+ yrs, female
  - Instagram – 17% of internet users, 18-49 yrs, urban, ethnic
  - Pinterest – 21% of internet users, 18-49 yrs, female, suburban
- Importance: High. 90% of information transmitted to the brain is visual, 40% of people respond better to visual information, photos enhance search engine optimization (SEO), and are the second most common form of sharing.[5]
- Usage note: Images should be high resolution, close-up, behind the scenes, very detailed – these are the ones that perform the best. Posed shots, low resolution, poor composition, long-distance, off-topic images do not perform well.

Video

- Example services: YouTube, Vimeo, Vine
- Features: visuals, audio, editing, storytelling
- Importance: High. Visual information is transmitted more easily, 12x more likely to be shared than links and text combined, viewers spend 100% more time on pages that have video[5] (see Resources, Media Bistro)
- Usage note: Videos that are 2-3 minutes in "long form" work best. Use editing software (YouTube has it's own limited editing features) to produce a clean video. No need to over-produce with formal sets. Good lighting, sound, and valuable content are the most important. Videos can be shot on low-cost equipment (mobile phones) but better results come from high-end equipment. Putting links back to websites into the captions garners high search engine optimization as well as driving traffic to desired locations. Google owns YouTube so it is favored in search results and reaches a wider audience than Vimeo though the quality of Vimeo far surpasses YouTube.

Social Networks

- Example services: Facebook, LinkedIn, Google+
- Features: offers ability to reach largest amount of people in the smallest amount of time, incorporates text, video, photos, sharing, and news
- Demographics: varies by service.
  - Facebook – 71% of US internet users, all ages but highest use is 18-29yrs
  - LinkedIn – 22% of users, 30-64yrs, college educated
  - Google+ – 320 million users, average age 28yrs, male, average user spends 7 minutes on site (very low)
- Importance: High. Facebook is the dominating social network and many people get the majority of their news on this site. LinkedIn has no competition in the business community but is a niche tool. Google+ is really important for search engine optimization, because the largest search engine owns it.
- Usage note· Facebook is becoming increasingly difficult place to stand out. Record numbers of

pages and users mean this channel requires a strong strategy to see benefit. Paid advertising is necessary to ensure visibility.

Bookmarking

- Example services: Pinterest, Reddit, StumbleUpon
- Features: post and save shareable web links into categories, boards, or lists based around images and videos
- Demographics: varies by service.
    - Pinterest – 21% of internet users, 18-49yrs, female (Pew Research Center)
- Importance: Medium. Best used for communication, resources, visual products, and retail.
- Usage Note: Pinterest is number two in referral traffic to Facebook's number 1. In first quarter of 2014, Pinterest was up from 4.79% to 7.10% while Facebook's referral traffic share grew from 15.44% to 21.25%. (Shareaholic)

As your usage grows more advanced or niche, you may find the following useful depending on your intended audience:

News and communities

- Example services: Twitter, Reddit, Newsvine
- Features: feature stories that are posted by users and some of these services allow stories to be ranked by others based on popularity. Posts can be commented upon and, in some cases, the comments are also open to ranking

Events and webinars

- Example services: Eventbrite, Evite, Cisco WebEx, GoToWebinar
- Features: designed to help you invite people to an event (whether physical or virtual), allow guests to share the event with others, and even implement ecommerce

Location-based services

- Example services: Swarm (spin off of Foursquare), Yelp
- Features: service that uses the global positioning system chip in mobile devices to provide information or entertainment based on the user's position on the planet, check-in function, loyalty rewards, gamification

A few tools you should know about that aren't social media by definition but you should know about them.

Link (URL) shorteners

- Example services: Bitly, TinyURL
- Features: a technique in which a URL may be made shorter yet still direct to a specific page on the web, trackable with analytics, brandable, identifiable, easy to share, works especially well with Twitter

Quick response codes (QR)

- Example services: open source, non-propriety. See Resources for sites that offer free creations. I like http://www.qrstuff.com and http://www.qrcode-monkey.com (that's how I inserted the Starter Kit logo into the QR codes throughout this book).
- Features: these codes have been used since the 1990's to track auto parts. When scanned with a freely obtainable mobile application, a user can be directed to a wide variety of websites from purchase points, video, basic text, or contact information. When left unbranded QR codes aren't

very attractive but are an easy way to convey a message to an audience. One must have an internet connection to use codes. Also, it doesn't make technical sense to include codes in places where the internet isn't available or on a piece of content that is most likely to be accessed in a mobile environment as the user won't be able to view and scan at the same time.

Ephemeral messaging applications

- Example services: Snapchat, Frankly, WeChat
- Features: private messaging services that give the sender the ability to time how long the message exists before disappearing and recall messages
- Usage note: Ephemeral messaging apps aren't truly social media as they don't exist on the internet and are by design limited in sharing. The privacy of these services is limited to the terms of use offered on each but much of the data is continuously stored on servers and can be hacked so the term "private" is false. I discuss the term "dark social" in Chapter 7: Measuring Performance and talk more about apps and services that exist outside of traditional social and measurement parameters.

The following services get into the data capture, advertising, and marketing functions of social media:

- Mobile marketing
- Search and social ads
- Display advertising
- Marketing apps
- Content marketing
- Cloud
- SEO
- Marketing data
- Analytics
- Business intelligence
- API's
- Testing and optimization
- Ecommerce

Quick Glance at Popular Tools

Facebook

Everyone is on Facebook. Think I'm exaggerating? As of summer 2014, it's at 1.32 billion monthly active users or MAU. It's one of the most common forms of digital communication, the largest social network, and easy to use. Most users access it from a mobile device (1.07 billion mobile MAU). The field is crowded and it's hard to be noticed. Advertising has become a necessity but the costs are quite low and targeting is exceptional.[6]

Twitter

Only 22% of U.S. internet users access Twitter.[7] It's fast, tough to "get", but is used extensively by two main audiences: urban and ethnic youth; and college-educated, higher income individuals with white-collar jobs. In the 6 years I've devoted myself to social media and technology training, this is the tool that proves the hardest to maintain and requires the most brainpower to master. Many people have a Twitter account but don't use it. I liken it to a blank canvas. We all start with the same white space but paint entirely different pieces. Don't use celebrities as your guide for recommended use as they tend to demonstrate the lowest common denominator of human behavior and thought.

YouTube

Everyone is watching YouTube. We consume over 6 billion hours of video per month. The younger the audience, the more heavily they use it. 18-34 year olds watch YouTube more than any cable network.[8] Users tend to watch more than one video in a row. This service has a low barrier to entry but requires planning in order to post regularly. Videos don't have to be professionally shot or expertly edited (though it helps) and should be around 2 minutes long. Users are more likely to watch 10 separate 2 minute videos rather than a single 20 minute clip. Additionally, it's the second largest search engine[9] and has the benefit of being owned by Google. This tool will only grow in importance.

Pinterest

In the U.S., this channel is dominated by women and the visual content they prefer though in the U.K. the majority reverses. It drives traffic extremely well but the users are coming from the fast paced/low attention span environment that is Pinterest so they don't stay on your website long. This is the best place to sell your retail products or handmade items. When you combine Pinterest with Etsy, amazing things can happen and sales can skyrocket.

Google+

This is Google's answer to Facebook. It has impressive design and superior tools (Circles-smart lists of contacts for information filtering, Communities-groups based on interest, Photos-high quality with editing and rotating carousels, and Hangouts-video chatting) but low user engagement. It's excellent for search engine optimization (SEO) but is a bit of an industry joke as far as true human interaction.

LinkedIn

Both personal profiles and company pages are beneficial but personal profiles garner more interaction and impressions. Company pages serve as a miniature duplication of a company website. I used to be a technical recruiter and if we could've built this tool, we would've. It's the best tool available for intelligence gathering.

I get asked questions about my favorite tool. I personally like using social media, it's why I got into the field. My favorite was Twitter long ago but now I say, "If this was

Survivor and I had to chuck everyone off the island, I keep LinkedIn." I think it's that important.

Instagram

Facebook bought Instagram in 2012 and this photo-sharing site has really taken off due to the cross channel exposure. In the now, behind the scenes, and very personal images are the standard. The personal and messy factor is frequently difficult for corporations to come to terms with. The platform is slowly changing as more businesses get on board. It currently has no desktop upload component so it must be used from a mobile device.

Email marketing

Regardless of the tool that delivers your emails to a large audience, this really is a must have if your audience isn't youth based as they don't prefer email. It's the most common digital communication tool and has the highest rate of conversion. Put a call to action in there, and they'll use it. These services have excellent analytics that stay fairly consistent for year-over-year tracking, which the other tools are poor at providing.

How to Determine Which Tool is Appropriate

There are a few important questions to ask when considering a tool for your campaign.

Is my audience using this tool?

Use Pew Internet Research Project to determine user demographics and engagement with a particular tool. Use a search engine to search for keywords such as "Pew Internet (name of tool) data" to locate information. This approach is often easier than searching Pew's site itself. Make sure you adjust your search settings with a specific date range. The results you will get from several years ago will be drastically different from that of this year or just the last few months.

How do people use it?

Research the tool and the ways it's used. The simplest approach is to ask a search engine questions such as, "What is (name of tool) used for?" "Why do people use (name of tool)?" Your goals and the demographics of your desired audience(s) should match for best results. Some of the cultural nuances stay the same but as time moves, changes in user behavior can happen.

Are others in my industry using this tool?

Investigate your industry colleagues and your competition to see what tools they are using and how. Note if your impression of their approach and messaging is favorable or unfavorable. Additionally, ask yourself which tools they seem to be missing. Are they not thinking outside of the box or is there a research-based reason for certain tools being omitted? If you decide to use a tool that your competitors aren't, you will be able to easily set yourself apart from the pack.

Will this service help my organization achieve its goals?

Many open accounts simply because it's perceived "everyone else" is on them. Have a good reason to delve into a new social media culture and make sure it serves your needs.

Other Tools You May Need

A sampling of other technology tools you will need to accomplish your goals:

- Desktop or laptop computer
- Digital camera
- Digital video camera
- External microphone
- Photo and video editing software
- Visual, animation, infographic software
- Cloud storage
- Google account for the company (not a person's account)
- Analytics (Google Analytics recommended)
- Management & scheduling tool (Hootsuite or Buffer recommended)

Beware of Abandoned Accounts

You will go through the process of researching social media platforms, what your audience is using, and determining which is right for your goals. You will set up the account, observe, interact, and measure. At some point you may determine that maybe that tool isn't right for your needs. That's okay. What I want to warn you about is abandoning your accounts.

When your customers look for you by name online, they will find your social media channels. They will find the ones you use and some that you don't use. When they take a look at your Twitter account (for example) that you haven't used since 2012, this does not leave a good impression. It's a good idea to remove or close down accounts that you are no longer using. It's better to clean up after yourself than to leave the web littered with bad choices.

You may not be able to access the accounts in question. Use Help or Tech Support available on each site to ask they the accounts be closed and give them the reason. By and large, they are helpful and understanding of your situation.

Best Practice: Just because everyone else is using Twitter, doesn't mean you should. Make sure the channel is appropriate for your goals, your audience uses or prefers it, and do your homework before deploying a service. Start one at a time so as not to overload yourself.

CHAPTER 4

Content Creation

Types of Content - When you begin to make decisions about what kind of content you'll be sharing, think about these seven basic uses of social media tools:

- Communicating
- Cause support
- Contests
- Consumer research
- Connecting with others
- Customer service
- Community building

Picking themes for months is a good way to begin your editorial calendar (see Resources for "15 Easy Content Topics" for theme examples). These topics are called "evergreen" because this content isn't time sensitive and can be used on an ongoing basis. These items can be bumped or moved to a new position on an editorial calendar if something with a higher priority arises.

Sample of ideas:

- Events
- Feature service partners
- Case studies or client success stories
- Testimonials
- Hot topics or newsworthy items
- Frequently asked questions

Different ways your content (or services) can be expressed:

- Interview
- Editorial
- Review of programs and services
- Definition of terms
- How to
- List (very popular when started with a numeral, such as "5 Ways to Get Healthier Today")
- Department, team, or team member profile
- Case study
- Comparison
- Parody
- Frequently asked questions, Q&A

- Special report
- Infographic (visual representation of data)
- Project
- Survey
- Contest
- Comics and jokes
- Presentation
- Podcast (digital audio recording)
- Webinar (digital video recording that is designed for teaching and conversation with audience)
- Video

Build a House of Social Media

Your social media campaign is like a house. The foundation and the base of it all is your website or other place where you convert visitors into buyers or users of your services. Each channel has its own theme, feel, culture, and rules. Every time you want to share a piece of content, you'll need to know in which "room" of your house it belongs. Many people assume it's about gaining likes or followers on social media. It's not. It's about what affects your bottom line. (see Resources for an empty house graphic)

Think of these tools as slippery slides that smooth the path for your prospective and current customers to spend said money. Social media isn't the point. It's a vehicle to offer your audience opportunities to learn more about you, spend money with you, or to gain knowledge from you.

Each room of your social media house serves a different function. Tools that duplicate functions, audience or theme may be overkill and unnecessary. I had this house visual created for my clients and students so that they can better see the environments that each of the majors is about. Read my descriptions that go with the picture:

Living Room | Facebook

Facebook was designed for college students. If you've ever been in a dorm, you've seen the common area. These rooms are filled with couches, ping-pong tables, TVs and people hanging out. These people are wasting time, visiting with their friends, eating snacks, and playing games. This is NOT the place to hard sell your audience. This is their living room.

Messaging tone: conversational, lightly informational (5th grade reading level)

Personal use: to talk to my clients and audience. I keep it light and Cliff's Note short. There are a lot of my target audiences present (yours are too) so I balance my content for businesses, parents, and individuals. Additionally, I post about keeping kids safe online because I know parents appreciate it.

Kitchen | Pinterest

This is a crafter's heaven. You can get endless ideas on food and recipes, crafts, art, photography, travel, fashion, shopping, and so much more. Pinterest is one of the best ways to sell your products and a favorite way for users to waste time. Like an after dinner snack, the activity here peaks at 9pm.

Messaging tone: visually arresting, informational, planning, bookmarking to save for later

Personal use: to share innovations and information that can be easily shared by category. I also use Pinterest to create resources centers that I can share with industry specific clients. Some of my boards: Tech & Social Media Info, Tech for Disabilities & Special Needs, Tech & Government, Technology in Education, and Great Tech Products.

Office | LinkedIn

This professional networking tool has no online competition. Busy professionals make connections here, express and declare their thought leadership, vet reputations, promote themselves, and investigate the competition. No puppies, no kittens, no bacon, or boring conversation. This is all business.

Messaging tone: informational, value-based, professional

Personal use: to share my own blog links to demonstrate my expertise, to promote upcoming classes, to share other pertinent technology or business information that I think professionals will find useful.

Bathroom | Instagram

This is the voyeur and the exhibitionist, a peek behind the curtain, the illusion deconstructed, the "candid" shot of ~~who we really are~~ who we want the world to see. We are all teenagers trapped in the bathroom trying on who we are. As I mentioned before, instant and raw, unpretty and unvarnished is what people expect here.

Messaging tone: fun, action-based, behind the scenes, close to the subject, artistic

Personal use: to show the personal side of my business and the world as I experience it. It's really the only place that I show my goofy and silly side.

Gym | Google+

My joke is that it's like the gym – everyone belongs but no one goes. It's quiet and sparse, good for your (SEO) health, but not much fun.

Messaging tone: informational, visually interesting, technical

Personal use: to share my own blog links (for increased SEO) as well as other geeky, hard core technology information I come across.

TV Room | YouTube

The TV of today has the most engaged traffic in the social world. This is a legion of people willing to watch "just one more."

Messaging tone: informational, humorous, musical, and educational

Personal use: I don't use it nearly enough. I tell stories about the kinds of questions I get from my clients and the public. I make videos to "tell" the story I think you need to hear.

Bedroom | Twitter

Gossip, rumor, and viral content travels at the speed of light. Twitter is the good, bad, and the ugly of human communication. Twitter is simply the most effective tool to spread information around the globe, nothing can beat its speed (a little like a teenager's ability to communicate). This is the one with the potential to burn it all down. Respect.

Messaging tone: conversational, educational, informational, thought provoking, fast

Personal use: to stay abreast of technology and world news. It's more important to me as a news source than a publishing platform. I look for trends and follow people that I look up to for their business acumen, innovative edge, or thought provoking information. I post links from others, my own blogs, promote my classes, and try to have conversations (that's where I get my best impressions).

When you determine the character of each channel ahead of time, it will help you decide what content belongs on where. When sharing content from location to location, you can alter the message to fit the tone for each channel.

What you need to remember is that your social media home will fall apart if the foundation isn't strong. You are driving traffic to your points of conversion whether that's an email sign up, a shopping cart, or other action you want your customers to take. Social media channels are shortcuts to the prize.

Excerpt from Keep Your Social Media Home Grounded, *a blog I wrote in May 2014.*

Editorial Calendar

It is recommended that you keep an editorial calendar (see Resources) for tracking and planning content creation and distribution. I've provided two versions; decide which you prefer to use as your starting document. The simple version was adapted from traditional marketing templates that feature print, radio, television, and press releases. Most likely your organization will continue to use traditional methods of marketing so your editorial calendar should include the plan for those as well. The second is adapted from a variety of sources to demonstrate how much more information can be collected and analyzed.

The construction of the editorial calendar isn't important as long as it serves the function of listing the tools you are using to distribute marketing messages and timing to be used. As a starting position, using months is recommended. As you (or your social media team) grow more confident, adding individual weeks will be of benefit.

The templates included in Resources are truly samples. They are both built in Excel. Create your own (or start with the populated template) in a spreadsheet of your choice and feel free to add, delete, and rearrange as required. I prefer to house this information digitally because it's really easy to edit and move things around without making your working document messy. I recommend including your Goals as a reminder and have your Audience identifiers listed as well. If you are working on this document with a team, it may be necessary to use a real-time collaborative document-editing tool such as Google Drive.

In larger organizations, the need to have your content approved ahead of time is common. Keeping an editorial calendar will make this process much easier. In this instance, your calendar can also serve as a record of posts.

The editorial calendar can be kept in a multi-tab spreadsheet environment, which I call a Master Social Media Document, and simplifies the workflow process.

Example of tab names within a master social media document (see Resources):

- Digital asset list
- Editorial calendar
- Timely task list (more in Chapter 5)
- Metric report (more in Chapter 7)

Best Practice: Know that the editorial calendar will change. You may not hit your targets for content creation and distribution. If a consistent pattern emerges where you are not being successful in following your own plan, an evaluation of obstacles are in order. Make changes where necessary.

Balancing Your Content

I love to sit in on social media training sessions taught by other instructors. I get inspired, I hear new ways to present material, and I learn something every time. I attended a session taught by Elyse Tager of Constant Contact several years ago and she said something that really stuck with me. She presented a formula to help you balance your content that you should know.

70% Value + 20% Promotion + 10% Fun =

100% Awesome

This is an exceptional guideline to use in your content creation (I added the awesome). Seventy percent of the time, your content should be useful information that your audience needs or wants. The helpful content you provide will show your audience that it's not about you all the time, you are givers (not takers), and concerned with their needs.

Ultimately, you should get a return on investment with your social media so promotion is necessary. Twenty percent is a good amount to get the job done. Use calls-to-action such as "Book Now" "Call for an Appointment" "Sign up" – action words to tell them what you want them to do.

The last part is the one that makes organizations feel awkward. Ten percent fun is actually necessary. People don't want to do business with logos or faceless organizations. We want to like the people we work with, we want to support those in our community, and we want to feel connected. You will have to show your humanity otherwise, you'll sound robotic and your audience won't identify or sympathize with you. Fun doesn't have to be crazy or ridiculous and it doesn't mean go "off the rails". Here are some examples of being real:

- Take your dog to work day
- The smiling faces of your staff
- Birthday party in the office
- Fun office jokes, practical jokes
- Staff barbecue
- Under construction progress and your adaptations
- How much everyone enjoys their job
- Getting together outside of work
- The extra skills or interests of people that work in your organization
- How you support the community and why

Where to Get Ideas for Content

One of the most difficult parts of creating content for your social media channels is learning where to get it. It takes practice but once you get the hang of it, it's literally everywhere. Here are some examples of places to go looking:

Staff Meetings

The needs of clients, the obstacles you are facing, the awards or funding you've received, new services, retired services, promotions, major changes, etc.

Memos

These are easy to save. If there's something good in one of them, stick it on your editorial calendar or copy it to your content receptacle.

Event Calendars

These tend to come out regularly and have lots of shareable and valuable information. Using scheduling tools, you can plug in lots of promotion for upcoming events in very little time. Holidays are excellent sources of events. These can be Federal, Bank, or fun holidays such as National Peanut Day.

Follow social media accounts of similar organizations

See what others in your industry are talking about online. They will talk about services similar to yours in different ways than you do. Observe how their audience engages with their content. Make note and take cues on what's working and what isn't. If you find one that is doing a great job, see how you can use similar tactics on your own channels.

<u>Subscribe to industry emails, periodicals, thought leaders</u>

Email newsletters are excellent for saving until later. Go to your favorite industry websites and see if they have an email signup box. When you receive these, you can save them to an email folder for batch reading or resharing. If they send too frequently or the information isn't useful, you can remove yourself from the mailing list easily.

The information you read in other places can be shared directly from websites and can be sourced in your content.

<u>Use a content curation tool</u>

Content discovery tools are useful because they allow you to monitor feeds based on keywords and trending posts/topics.

Follow The Rock Star

Smart leaders pay attention to the successes of the leaders before them. Even smarter leaders pay attention to their successes **as well** as their failures. You didn't invent the wheel. Many have come before you and have done great and terrible things long before you even thought of it, whatever it is.

In social media, there is a convergence of marketers, business owners, communication specialists, tech geeks, marketers, PR people and then some. We are all swimming in the same pool, learning from each other. Nobody gets it all right all of the time. We will all fail (and if we don't we're not really in it). The trick is to learn the biggest mistakes from those that made them before you rather than stepping in the muck yourself. I call it "following the rock star".

Let's say you sell toothbrushes. Are you aware of Colgate, Oral B, and Sonicare? Of course you are. Would you check out their website, blog, and marketing materials occasionally to know what they are up to? Most likely, if you are looking to be successful that's exactly what you are doing. You want to know what has worked for those that are already successful. And knowing what our competition is doing is vital part of doing business.

Now extend that one step further. Not only do you sell toothbrushes but you might be interested in dentists, denture makers, mouthwash, surgery centers, packing manufacturers as well as toothpaste companies. If you are paying attention to what these companies are doing, you'll start to see some "outside the box" thinking and tactics. Some record videos, others interview clients, some are just straight inspiring with the ways they have represented their business and services. Doing this kind of research will inspire you to get out of your rut and think differently about social media, I promise. When they do it **wrong** and when they do it **right**, you are noticing. This is "following the rock star". Whether they stumble or knock it out of the park, pay attention to your reaction as an observer (or prospective customer) and take that as a qualitative survey result. Leapfrog yourself over those stumbling baby steps and learn by example. Follow a rock star and you'll never regret it.

Here are some of my rock stars over the years: Mari Smith for Facebook, Cory Booker (former Mayor of Newark now Senator of New Jersey) for Twitter use in government, and Viveka von Rosen for LinkedIn.

Follow the Rock Star is a blog I wrote in January 2010.

Team Communication

There will be more than one person in your organization that will have information to share via social media channels. Regular staff meetings can be a great time to get this information. This can be added as an agenda item taking 5-10 minutes of staff time. Ask everyone what they feel is important to be distributed on social media and what they think the clients need or would want to know. Many items will be time sensitive and need to be placed on the editorial calendar with enough time to realistically promote.

It is recommended to establish a regular communication format with your team. This may be in the form of a once a week email from the social media manager to appropriate staff requesting or prompting for the same information as above. A statement such as, "We work on our Facebook/LinkedIn posts on (a specific day). If you don't have anything you'd like to be posted, no response is fine. If you do want something posted, it must be received on (another specific day)." Use a folder within your email structure to set aside those emails for when it's time to add to the editorial calendar. Additionally, the information may be copied and pasted from emails or other digital communication into a timely calendar update.

Content Receptacle

Identify a location for shared content. You may not be the only person in charge of the content that has been assigned to the editorial calendar and you may be collecting a lot of information for use at a later time. It's necessary to store what you are collecting in a designated spot such as a folder on your shared network, in Google Drive, Dropbox or some other similar location.

This is an excellent location for sets of film that are downloaded from digital cameras, interdepartmental memos, emails, or any other type of valuable shareable content that you don't need immediately.

Public Relations Assets

It is important to have access to your organization's approved logos, graphics, images, branding, color palettes, branding colors (on the web they are called hex #s and web safe #s, ask your graphic designer for these), boilerplate text, press releases, media contacts, protocols and any other assets you use to promote. It's also recommended to include in your Master Social Media Document a list of media contacts appropriate to your industry. Also include information on how to reach your IT department for website updates.

Establish Vendor Partners

Identify partners with whom you can create a cross-promotional team. These are vendors that would make sense to team up with so that you can both benefit from the relationship. Hairstylists and makeup artists, painters and floor installers, accountants and lawyers are just a few examples.

Reasons to team up:

- Your audiences or goals may overlap
- That vendor's information is valuable to your audience
- The two (or three) of you are currently partners (or have been in the past) on a project or objective

Create a line of communication between both (or multiple) organizations so that on a timely basis, you share information that can be included in your editorial calendar. This reduces the amount of content your team will need to create.

In social media, there is much emphasis on creating content but not enough is put on sharing the content of others. Every social media site has their own word for it but the concept of resharing the valuable information that you or your team members see in other places is important and lends credibility to your own messaging. Your audiences want to hear other voices that share tone, message, mission, and intent with yours.

Helpful Tools for Content Creation

There are literally thousands of tools out in the world for you to use that make content creation easier on you. Listed/linked below are a few I think you should know about. These are all free to use or download though you can pay for advanced features, I'm sure. If you find you don't like these or enough time has passed since I wrote this, you can always use a search engine to type in "best collage maker" "best meme generator" for example to get more options.

- Slideshare is one of my favorites. I call it YouTube for slideshows. It's so good it was purchased by LinkedIn in 2012. Its main function is to repurpose your slideshows whether they are Powerpoint, Keynote, .pdf, or OpenDocument. You can include audio and video. Slideshare offers great search engine optimization and is a great way to share what your value is in a "digital brochure" kind of way.
- PicMonkey / Canva / PhotoFunia are all amazing photo manipulation tools. Overlay images, text,

create collages, templates for all social media header images as well as profile and post images, and they all have amazing effects.

- Eas.ly is a great graphics and infographics creator full of templates.
- PicStitch is my mobile collage creator.
- Fotor is a collage maker for my desktop.
- Videolicious is a great mobile video recording and editing tool.
- Producer is a mobile app for creating your own memes (those funny photos with block text laid over the top) also called a meme generator.
- Flipagram turns still images into a slide show with choices of music.
- Skitch is the tool I use on my desktop for marking up screenshots.

Content Use Cycle

The content you create, whether it's a written blog post, status update, image, video, or graphic, can and should be used in multiple places. A status update that garners a lot of engagement can be turned into a longer blog post. A traditional press release can become a blog post then a two-line status update. See the visualization below to get an idea of the flow of your content. You won't need take every one of these steps, use your discretion about what will and won't work for a particular creation. The reason

Blog is sitting at 12 o'clock is because it's the most important (see Chapter 3: Tools for more on it's importance).

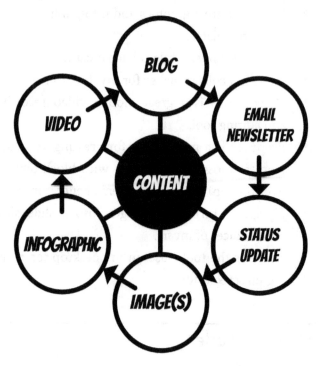

To reduce the amount of work required to manage multiple accounts and to maximize visibility for your message, you will want to share your own content from channel to channel.

Example of what my process looks like: I have **email newsletter** on my calendar for the 20th of every month. If that date falls on a weekend or a Tuesday, I move it to a better day. (My audience doesn't want to read my material on the weekends and I only know this from years of watching their reactions to my content on different days at different times.) As I prepare to write the **newsletter**, I think about what my clients are asking me or what's a hot topic and then I write a **blog** on the subject. That **blog post** becomes the main article in my **newsletter** and links back to my website. Often I create a **status update** to share the **blog** link. (Commonly this whole process starts with the status update and then becomes a blog.) I will find or create a visual **image or infographic** to express the same concept (if it makes sense) and share on appropriate channels. Regularly I shoot **videos** that are embedded into the **blog** and when a new user finds my video, the caption contains the link to the matching blog.

Best Practice: Make sure you don't share identical content across multiple locations at the same time. If a member of your audience sees exactly the same post multiple times, they will cease to follow your accounts where the duplication occurs. Make sure you include a call-to-action (CTA) in your posts. This CTA will vary from channel to channel and will support your goals.

SEO and Being Found Online

SEO stands for search engine optimization. SEO is the process of affecting the visibility of a website or other online channel in a search engine's natural (unpaid, also known as organic) search results.[10] You've already identified important keywords that are descriptive to your products, benefits, value proposition, and services. Your audience will be searching online using those same keywords so make sure to identify and use them regularly. Use them in about pages, info sections, bios, captions, and regular posts.

Writing with SEO in mind isn't the only way to get found online. Make sure that your clients know you have social media channels. Use a variety of ways to inform your clients such as posters at places of service, list them on your website, put into email signatures, mention it in conversation, cross link your social media channels to each other, as well as include the information in handouts and other collateral. It's also important to highlight the benefits of the channels and value the audience will receive. "Follow us on Facebook" isn't good enough. They will need to know **why** as well as what's in it for them.

If you want to know how to get on the first page of a search engine, you've got two big ways to do that. First, in the design of your website, your keywords and basic structure will be built into the site by the team or company you contracted to build it. The second method is to use social media and directory listings. Much of SEO today is actually elbow grease and is sometimes called SMO or social media optimization. The effort you put into creating content on the web that supports that you know what you're talking about, are an expert in your industry, and that you are a regular contributor to the web in general are extremely important to being found. Buying advertising to assure you a place on a search engine isn't SEO, it's paid advertising, and that's also a good idea. I don't specialize in advertising so I won't advise in this arena.

Images and Copyright

It is necessary to incorporate images into content and it's important to have the rights to use all images and video. The best option is to create your own works or use that which you've purchased with reproducible rights. When using the work of others, you must know the copyright and rights to use. See Resources for royalty-free and public domain image sources.

Using your editorial calendar to plan ahead is a great way to make sure you have the images you need as well as the rights to use them. Set up photo shoots with staff and clients that sign a photo release (see Resources). Use props or feature events that demonstrate your products,

rvices, or value proposition. If you think like a photog-
rapher and set up your own photo shoots with staff using
simple digital cameras or by hiring professionals, you will
have a store of your own photos in your content receptacle that you can use again and again.

Using stock images is an option, if a budget is available
(see Resources). Use these sparingly, if possible, for your
audience knows stock images from shots of real situations and stock images don't induce trust. For hard to
achieve shots, those with artistic elements, or an image
that is particularly valuable, paying for them is often necessary. Picking and choosing carefully about which you
will pay for and buying images that are going to get a lot
of use are smart buys.

CHAPTER 5

Timing, Scheduling, and Automation

How frequently should you post? - There is no easy answer to this question as each channel is different. Researching each channel before using it is recommended. You will want to understand who is there, what the culture is like, and how often that average users post. Both Mashable and Pew Internet Research are excellent sources for this information (see Resources).

A fast list of timing recommendations for the most popular channels listed as minimum times per timeframe and maximum times per timeframe. Of course, you may not be using all of these:

- Facebook – min 3x/week, max 3x/day
- Twitter – min 1x/day, max 11x/day
- Instagram – min 3x/week, max 3x/day
- Pinterest – min 1x/week, max 6+x/day
- Blog – min 1x/month, max 2x day
- Google+- min 1x/week, max 2-3x/day
- YouTube – min 1x/month, max 1x/day
- Email – min 1x/month, max 4x/month (only retail & food can go this high)

You may think these numbers are impossible; they are averages, but it's not necessary to achieve this level of activity. Work closer to the high end, you won't be starting there.

Best Practice: Research each channel before using. Monitor and observe behavior regarding frequency before posting. Deploy one channel at a time to avoid overload.

Scheduling Tools

There are several ways that you can become more efficient in posting (it's one of the only ways a marketer can scale). There are third party scheduling tools as well as scheduling functions built into some of the services.

Hootsuite (https://hootsuite.com)*

- This social media management service can schedule posts into the future on a variety of channels such as Twitter, Facebook personal, Facebook groups, Facebook pages**, LinkedIn personal, LinkedIn groups, LinkedIn company pages, Myspace, Google+ company pages, Foursquare, and Wordpress.com.
- It can autoschedule your posts for optimum timing.
- You can upload an Excel or .csv spreadsheet for bulk post scheduling.
- Also features saving searches, monitoring connected channels, and access to analytics (paid accounts). Analytics offering are constantly being upgraded.
- Paid and free options are available.

I prefer Hootsuite over Buffer which is known for glitches and missed or repeated posts.

**Facebook has been known to suppress posts that are originated from a third party source such as Hootsuite. It is recommended to use the native scheduler listed below for maximum visibility.*

Buffer (http://bufferapp.com)

- This social media management service can schedule posts into the future on a variety of channels such as Twitter, Facebook personal, Facebook groups, Facebook pages, LinkedIn personal, LinkedIn groups, LinkedIn company pages, App.net, and Google+ company pages.
- Includes analytics based on the shortened links it creates.
- Paid and free options are available.

Facebook scheduler

- When clicking inside the status update box, a page will have a clock in the bottom left hand corner of the box. Creating a post then clicking on the clock will give the admin options for backdating or scheduling posts into the future.
- Depending on the age of your page, some must have a "start date" other than "Joined Facebook" in the About/Info section in order for this clock to be activated.
- This is a free internal feature on all pages.
- It's best to schedule posts to Facebook through their own scheduler. They have been known to suppress the visibility of posts that are delivered through third party tools.

Twitter (https://ads.twitter.com)

- A new tweet box is featured in the upper right hand corner (this button may move) and from this module the tweet can have targeted delivery, promoted (paid advertising), or be scheduled (for free).

Dlvr.it (http://www.dlvr.it)

- Using this tool, one can schedule or promote content in addition to distributing published content to be auto posted to others. Includes blogs, Twitter, LinkedIn, Google +, Facebook, Tumblr and more.
- Includes analytics.
- Paid and free options are available.

Benefits and Dangers of Scheduling

The number one concern about social media account management is about the time and labor necessary. Scheduling content ahead of time allows batch processing and maximum efficiency. The admin makes strategic decisions about channels, timing, and tone all from a central location. It increases your ability to repeat post the same valuable information multiple times and locations by simply copying and pasting.

Suggestions for use:

- The same post can go out at the same time once per week, month, season etc.
- Admins can promote events in multiple ways over periods of time in just one sitting.
- Admins can schedule posts to come out in the evening, holidays, weekends, vacations, or any other desired time.

The dangers of using scheduling and auto posting aren't always apparent. They can present in the following situations:

- Scheduled posts that reflect they were written on another time or date. Mentioning times of day or weather that shows the poster isn't "present" especially if unusual weather is occurring such as a rainy day in the middle of summer.
- Scheduled posts that fall on days of a violent conflict, crime, or other tragedy. The change in context can take on unexpected meanings.

- Scheduling and auto posting 100% of an account posts and thinking that substitutes for activity. If it's all canned, the audience knows.
- Something goes wrong with the scheduling service and posts come out incorrectly, with broken links, at the wrong time, or not at all. This is quite common.

Best Practice: Use scheduling with care. Monitor upcoming posts to make sure they are still appropriate. Monitor your account, read your own feed, and test out your own links within posts to ensure quality control. Be present (as in, don't schedule 100% of your posts). Be quick to respond.

Automating Posts

Automation occurs when using a linking tool to automatically share real-time content from one location to another. I like to call this "daisy-chaining." It's a favorite of people attempting to save time. Personally, I'd rather take the small amount of extra time to appropriately place my content to allow for slight change in tone, length, etc., and to stagger delivery times for maximum impact.

Common uses:

- Facebook posts pushing to Twitter
- Foursquare to Twitter
- Tumblr to Twitter
- Blogs to LinkedIn
- YouTube posting to Google+

This type of linkage rarely works the way an admin intends. It also irritates your audience. They know you aren't actually on the channel and it comes across as if you don't care.

Best Practice: Use automating processes with caution.

Labor Hours

The amount of time one **could** spend managing social media is endless. The amount of time one **should** actually spend on it is much more limited. I've seen several books on the market with titles similar to *Facebook Marketing in An Hour a Day!* My first thought is "What about the other channels? Am I supposed to spend an hour a day on each?" No one has that kind of time.

As a result, my goal has been to provide my clients with a plan for effectively managing social media in approximately an hour a day. See an example of how time can be mapped out below using the Timely Task List (also in Resources).

<u>Daily</u>

- Check messages. (5 min, 2x per day)
  - Check notification emails in main email for each account. See settings for frequency and type of notifications. Navigate to channels regularly as the email notifications are not reliable. Opening the accounts and navigating to them is the surest way to catch all messages from your audience.
  - Make bookmarks of your channels on your favorites/bookmark bar for faster navigation.
  - Using mobile applications is one of my favorite ways to check messages quickly. They have handy little notification bubbles that make it easier. Within one minute, I'm able to open phone, email, Facebook personal, Facebook professional, Twitter, Google+, and Instagram to check for messages. I do this at the start of the day, between meetings, at end of day.
- Engage on one channel each day (20-30 min+)
  - Assign yourself one channel per day such as Monday is for blogs, Tuesday is for LinkedIn. Navigate to the channel, read your stream, check your notifications,

follow new people, interact with other accounts, ask questions and seek answers, post and/or schedule content.

o The purpose is to "be there". These channels change quickly so if you only use them every once in awhile, they look foreign when you visit them and it makes you less likely to use them.

Monthly

- Metric report (20 min)
- Email blast (2 hours)
- Blog writing (1 hour)
- Batch post scheduling (.5 hours)
- Shoot/post videos (based on quantity to shoot)

Best Practice: Pick one channel per day for prolonged attention and rotate.

CHAPTER 6

Measuring Performance

ROI and How to Tell if It's Working - The only way to know if your social media actions are working is by analyzing your results. Previously you set SMART goals with measurable benchmarks and now we will use those to determine your return on investment (ROI). There are a lot of formulas out there that you can use to tell you how well you are doing but I like one of the simplest:

Formula:

(Benefits – Costs) / Costs = Social Media ROI

Example:

14,000 – 2,000 / 2,000 = 6 or 600% ROI

You are doing well when you are increasing your returns while your investments are staying the same. You are not doing well when you are spending more in labor, materials, and other investments than bringing customers as a result of using social media. The return is tied to your goals and you will be the one to assign dollar value for the benefits for this formula.

I've collected a few examples of ways to assign value to your goals, more advanced formulas, and all around much more information on this subject are available in Resources. Data analysis is the new "it" industry and is a huge area of growth because today's businesses and advertisers have a voracious appetite for data. Plain and simple, we need data to help us make smart decisions.

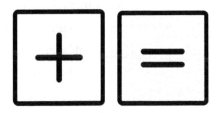

Your social media channels are like children (trust me, it'll make sense in a moment). You see them every day and because of that you can't see them grow. This is why we measure their growth with a mark on the wall every year. (See? I told you I'd get there.)

Measure your performance using ROI formulas and monitoring engagement to prove over time that what you are doing is working or not working, what your audience likes and hates, what cuts through the noise and what they couldn't care less about. Once you start using the channels and watching the reactions, it's quite scientific. Adam Savage of Mythbusters has a saying,

"Remember, kids, the difference between science and screwing around is writing it down."

Write it down and you'll see patterns of your behavior that affect the behavior of the audience. It's your turn to be MacGyver and Sherlock Holmes wrapped into one. There are no perfect answers to how you should run your social media campaign and some of the fun is being presented with obstacles and coming up with crafty solutions. This is the time for you to take ownership over the process. You make decisions about goals, channels, messaging, and timing all along the way. It's not until the measurement phase that you get to find out if you're brilliant or just so-so. It's like getting a present with the suspense of not knowing what's in the box.

How to Build a Metric Report

A metric report is a document for data collection of your social media performance and is best used to measure audience engagement. Two spreadsheets are available (see Resources) to start the process of collecting data. The simple template is clearly easier to use than the advanced. Use the one you think best fits your skill level or needs.

Basic features that a metric report may have are:

- List of tools used
- Metrics categories (see Channel Analytics below)
- Goals
- Timeframe
- Conversion due to calls-to-action

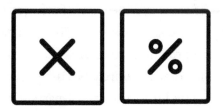

Best Practice: As mentioned in Chapter 4, the master social media document can be kept in a multi-tab spreadsheet that contains your digital assets list, editorial calendar, timely task list, and metric report. Keeping it all together simplifies the workflow process.

Channel Analytics

I recommend doing your data collection at the beginning of the month, the very first day, if possible. I worked in accounting for a few years and I find tremendous value in the wrapping up of a month before moving on to another. Build it into your routine and perform at the same time as close of month, inventory, and other important tasks. I like the routine of it. One final reason is some channels only offer "Last 30 Days" as a report rather than allow you to alter the date range. In this case, it's important to get that report as close as possible to the beginning of the month so your month-over-month comparisons don't have any overlapping days.

The process of collecting the data is fairly quick. Navigate to the individual channels and record the analytics they provide. Some provide intense amounts of information, some just a bit, some are unreliable, and some offer none at all. You may find that using third-party tools is desired to enhance the information you gather on your performance. It's very likely that you will be overwhelmed with data. As you investigate and learn what the definitions are regarding data reported, you will better know what is important to capture. The simple template lists most important information necessary to understand performance. As you grow more comfortable with the process, you can increase the amounts of data you collect to tell a more nuanced story.

My favorite benefit of using the metric report to record analytics is that by design, I must visit every channel natively at least once a month. This is what I exclaim every single time, "Whoa! Something changed!" I login to the service, take a look at my analytics, and then navigate to my own profile. I make decisions about whether or not I need to update a profile bio or photo and ascertain if it still looks and works the way it did last time I was there. New features are deployed and old ones that I love disappear. I take this opportunity to look at my posts. I use scheduling in some cases and this is the time to double check that the content is being published as I intended, with working links, and that my balance of communications is good.

Best Practice: Remember the 70/20/10 rule to gauge balance of content – **70% value / 20% promotion / 10% fun**

Invariably, changes and updates need to be made to accounts so I add these items to my to-do list. Now that I've been doing this for many years, I tend to take care of issues right there on the spot. If I wait, I may not complete the task.

As in all social media, the analytics offered to you will change over time. Frequently, the data isn't available in the exact same format so that you can accurately compare year-over-year as the services rename, combine, or eliminate categories. Quarter-over-quarter records are going to be easier to use to create a long view of performance.

Warning: There are at least two pitfalls of data analysis of which you should be aware. The first is what I call "geeking out on data." This resembles a black pit and you can get lost in the reams of information provided. Get in, find what you need, learn a little more every time you go, but don't get lost! Much time will be consumed here.

The second concern is when you let each fluctuation of data change your direction or goals. It's easy to doubt yourself and second-guess every move you make which can render you frozen and afraid to make any choice for fear each is wrong. (I call this "data paralysis") Now if your performance consistently suffers, then you'll need to reevaluate but every dip isn't a time to jump ship or make drastic changes. Some months are good and some are bad. Don't let the numbers run you. This is yet another reason to only check your analytics monthly rather than daily.

Reporting and Review

Every organization has different protocols on reporting return on investment of marketing efforts while some have none. Where social media reporting or management begins is often the Marketing department and their larger picture.

The numbers collected in the metric report don't say much by themselves. You will need to interpret what the numbers signify. Some factors to think about: increase or decrease in reach or fan base due to a particular action; advertisements you've run; changes in wording; new channel added to the mix, different timing; and changes to the sites themselves. As you continue this practice, you will improve in your ability to "tell the story behind the numbers". It's hard to get a complete picture until you have at least a quarter's worth of data. Month-over-month can contain a fluke that distorts comprehension but quarter-over-quarter allows you to spot and report patters. This also helps in justifying your investment in social media as to how it affects sales and other financial and service-based goals.

If you are an admin to these accounts, you will have a more intimate relationship with the fine nuances of the data. When reporting to others, their lack of knowledge in this area will hinder their understanding. It is recommended to provide colorful graphs, simplified bottom lines, and easy to digest snapshots to highlight the important information. Also provide detailed metric reports and other required reporting at the same time. Everyone involved can choose the version of reporting that works best for their comprehension. This two-pronged approach work best for audiences or teams with a spectrum of technical ability to understand goals, actions, and what it means to your department or organization.

Record Keeping

Some organizations, such as governmental bodies, have in their communication policies that it is required to keep a record of social media posts. It's an evolving field but here are several ways to keep records of social media activity.

- Use your editorial calendar as a diary of activity for posts.

- Use the download activity, if the service offers such an option. Many of them don't offer a reliable way to capture activity between an admin and the public.
- Create screen captures of activity. Learn how to take a screenshot or capture from your computer. Save in content receptacle.

If your company has the budget or deems it necessary, these tools are all paid and perform social media account backups:

- Frostbox (http://www.frostbox.com)
- Backupify (https://www.backupify.com)
- SocialSafe (http://www.socialsafe.net)

Best Practice: When discussing social media management with your team, clarify how this new reporting will fit into current procedures and preferences for delivery and frequency of reporting.

Dark Social and The Immeasurable

There's a cool term coined by Alexis C. Madrigal[11] at The Atlantic that makes me think of superhero villains–"dark social"–though it's not as nefarious as it sounds. It refers to social traffic that occurs on the web that's largely invisible to most analytics and reporting programs. As people use the web, the ways in which they arrive at websites are tracked but some traffic slips through the cracks. Types of illusive traffic: direct links, when they are saved as bookmarks, when clicked from within emails, from instant messages, and some mobile usage.

Many in the business of marketing don't talk about the fact that the data we get isn't always accurate or even the whole picture. You are never going to get a perfect reporting scenario as some factors are unknown. This isn't new. We've seen this with traditional advertising tracking as well. Arbitron is "a leading research company known for producing radio and television ratings"[12] and they've had quite a few missteps in calculating the viewership of urban markets.[13] Nielsen Company[14] which measures consumer behavior uses self reporting diaries during sweeps for television ratings and this leaves the data collection up to those that are participating and isn't very technical. In 2010, it admitted an error[15] in data reporting that lasted at least one quarter and as a result recorded a 22% drop in web traffic. It ended up being a minor error in programming on Nielsen's part but it impacted at least five other large services in addition to custom research.

The reason I brought up errors and unknown factors is to point out that there are gray areas in all fields of data collection. There are a lot of opinions about the effectiveness of social media and the analysis around its performance. What I do know is that we have more data than we've ever had before, including when we used traditional media such as radio, print, and television exclusively. We have the data to make smarter decisions to target our ad dollars and time more efficiently than ever. Make use of all that data.

CHAPTER 7

Policies, Crisis, and Reputation Management

The importance of having a social media policy - I am frequently asked about social media policies and how to implement them. I want to direct you to your pre-existing employee handbook. Do you have one? If you don't, it's time to create one. If you do have one, your social media policy will live in the communications section. You may have ideas about how your employees answer the phone, what is included in their email signature, how to schedule webinars, the best procedure for storing sensitive information such as financials as well as passwords, etc. Social media will become as common as faxing was 10 years ago but because so many consequences of poor social media use can be damaging to the company's reputation, create vulnerability to lawsuits, and because basic education in the subject is hard to come by, a policy is necessary for many organizations.

There is no one answer for how to create a policy for your organization but I do have a great resource for you. Social Media Governance has a large index of policies from large corporations and all types of industries. Take a look at what they have and find a template that suits your needs (see Resources).

Listening

Much emphasis is put on the "talking" in social media but not enough is on the "listening." When spending time on each of your channels, you can see what others are saying about your program, needs that you serve, or any other subject deemed important to your goals. Active listening can give your team a heads-up when a contrary opinion or backlash is brewing. Additionally, you will see opportunities to offer assistance and gain new customers by providing proactive customer service.

How to Deal With Negative Online Feedback

Negative online comments are a fact of life and there's simply no avoiding that some people are going to be unhappy. It doesn't have to deter you from promoting your program and providing great service.

- <u>Listen</u>
  - o Make sure you are aware of what your clients are saying about your program. Bad reviews can highlight where larger issues are hiding.

- <u>Acknowledge the complaint</u>
  - o Everyone wants to be validated. Being polite goes a long way. Sometimes you will hear from people that have very few other things going on in their life and your kind response can diffuse a situation that if ignored, could spiral out of control.

- <u>Fix it fast</u>
  - o Your audience understands how much power they have online. Words go viral very quickly. After the initial contact with the client, let them know when a full response will be forthcoming.

- Take it offline
    - o Offer to communicate with them in a more private setting such as email or over the phone so that you can take it off the public stage. If the complaint is in a public arena, strive to take it private as soon as possible. It will be easier on the client and keeps further negativity out of the public eye.
    - o When a complaint originates in a public forum, make sure you let the public know on that same forum (and other place, if necessary) what the resolution is, that you care about customer service issues, and that you work hard to make it right.

- Be honest
    - o Care about your clients, engage when they want to hear from you, and offer to fix problems. Provide great customer service.

- Don't delete a comment just because it's negative
    - o If the public comment does not violate a stated public comment policy then it is good form to leave it up even if it doesn't

agree with the company stance on an issue.

- o Deleting posts damages credibility, can further anger the original commenter, and may alert your audience that something is amiss.

- Don't take it personally
  - o Unhappy clients happen. It's part of working with the public. You need to make sure you did everything you needed to do.

- Make changes where necessary
  - o By having your process in place you will be able to make necessary changes to the way you deliver customer service and improve the client experience.

Taking Responsibility

No one is perfect and we all make mistakes. It's important to acknowledge when you make an error because the chances of your audience finding out are pretty good anyway. When you take the offense and provide transparency, there's less opportunity for dramatic negative fallout.

Apologize if necessary. This can be hard for people, but sometimes the complainer just wants to be validated or acknowledged. This accomplishes two things: the customers are happier because they've been acknowledged and the general public can witness how you handled the situation.

Best Practice: Bad stuff happens. Listen and respond accordingly. Own up to your errors. The only thing you truly control is how you respond to a situation and it's how you react to a situation that the public remembers.

Common User Question

I can't count the times I've been asked, "Why can't I just turn off the commenting feature?" See my many answers here:

- The word "social" in social media is about conversation. If you shut off the other person's ability to talk, it's simply a bulletin board.
- When you have a conversation with someone in real time, having your say, then ignoring the other person is considered rude. Your audience feels slighted and upset when you ignore them.
- Your audience will not respect you for attempting social but not being able to cope with the good, bad, and the ugly.

In fact, this is so frowned upon, Facebook has almost completely removed the ability to shut down comments on their business pages. In 2011, Versace had its Facebook page besieged with public comments over their sandblasting jeans technique. They deleted many comments then shut down the commenting feature altogether. It blew up in their face, stained their reputation, and eventually Versace discontinued sandblasting.

Why Deleting Negative Posts is a Bad Idea

It's generally considered bad form to delete a post from your social media channels just because you don't like what that person has to say. As business owners and managers, we have to come to terms with the fact that the second we open our "doors," someone somewhere is unhappy. We can't please everyone all of the time. The **whole purpose** of using social media is to have conversations and communicate with others. If you are a page administrator who removes a post by the public simply because the content isn't what you prefer, then you don't understand what social media is really about.

I recommend deleting and removing posts from others if they are racist, sexist, full of hate speech, obscene, or violate stated community guidelines. Deleting simply because you don't like the content shows immaturity, or an inability to deal with real life situations, and damages your credibility. No matter how much you try, you can't remove all negativity from your world. Instead of pulling out the big pink eraser, acknowledge the concern (if they aren't delusional), communicate with the person, validate their concern, then discuss your plan of action, whatever it is. Remember, there is always someone watching your actions and there are silent members of your audience who will notice.

The way you deal with unhappy or negative people is proof of your character. A less-than-rosy comment doesn't have to be the end of the world. It can be a learning experience if you are open to it.

Have you ever had cruddy customer service, complained, then received excellent treatment and it changed the way you thought of the company? It happens to me all the time. People love to bag on telephone reps. I love it when I get a truly helpful and nice person on the phone. It happens more often than people acknowledge but sometimes it's the way Ginny from Oklahoma treats you that determines how you feel about the multi-billion dollar conglomerate. Take every opportunity as a chance to provide a stellar experience. It's never too late to turn it around!

Bottom line? Deleting posts damages your credibility. Are you wondering if you post something bad on my Facebook page whether or not I leave it? Go ahead, test me: http://www.facebook.com/KerryRegoConsulting

Be Transparent: Why Deleting Negative Posts is a Bad Idea[16] *is a blog I wrote in 2012.*

Creating a Crisis Plan

A crisis plan is vital to have on hand should your online assets be hijacked or your organization suffers other public relations situations. Treat it like an emergency services department would: create an "evacuation plan" and run it like a fire drill.

- Sit down with your team and brainstorm every possible crisis scenario such as hosting problems, hijacked accounts, denial-of-service attacks, or loss of access. Come up with a plan of action for each scenario.
- Have statements in place to release to clients, employees, and journalists over social media, your website, and your front door (if it needs to be closed) informing them of what is happening. Designate a spokesperson.
- Create an emergency list of contact information so management can be reached 24/7.
- Make sure you know how to reach the first responders in your vicinity and develop relationships.
- Once the crisis is over, reach out to your clients and employees to thank them for their patience.
- Perform test runs or fire drills to make sure all on your team know what they should be doing and any problems can be rooted out before a crisis occurs.

- Keep your crisis plan with your master social media document and also with other emergency information.

"The web is like your Hollywood agent:

It speaks for you whenever you're not around to comment."

Chris Brogan & Julien Smith, Authors of Trust Agents[17].

How to Manage Your Online Reputation

My first book, What You Don't Know About Social Media CAN Hurt You: Take Control of Your Online Reputation published in 2012, was entirely devoted to this subject. I sat down in the winter of 2011 and categorized every blog that I'd written up until that point and identified five important themes I wanted to write books about. The Social Media Starter Kit was originally going to be first but I felt the people who didn't want to use social media or those who were too intimidated by it were in greater need.

Your reputation can so easily be damaged without you even knowing it that there are some action steps I want you to know so that you aren't caught unprepared. Below is a blog I wrote in 2012 titled How to Get Started in Reputation Management.[18]

What is the importance of your online reputation? In a 2010 study by Microsoft and Cross Tab Market Research, 70% of U.S. recruiters have rejected candidates based on their online reputation though only 7% of Americans believe their online reputation affects their job search.

So you've decided it's time to get proactive in your reputation management. How do you start?

Perform a Vanity Search

You will need to monitor what is already being said about you so you know what other people are already able to learn about you and your brand. Do a vanity search for your name, business name, or known as names. Make sure you look up your founder/owner's name(s), business name, and industry plus location on Google, Bing, Yahoo! and any other search engine you prefer.

Own your own domain

Buy your name, variations, and business name(s) if you can. You can have them point where you want. I own

KerryRego.com and it points to my site KerryRegoConsulting.com. Use any domain purchasing service you prefer and buy it. They generally run anywhere from $3-$10 per year with a price break if you buy multiple years or domains at a time. If you are planning on using it to create content rather than simply owning it, purchase more than one year at a time. Search engines can see how long the domain has been purchased for and interpret that you are in business for the long haul. It's a ranking factor that pushes you up higher in search results.

True story: I have a friend that is a Broadway performer and she didn't buy her own domain before someone else did. When one does a search for her name a XXX performer comes up before her. **Own your own name online.**

Post original content

Don't be passive, be proactive! Get yourself a blog (some are free) like Blogger, WordPress, Tumblr. Decide on a focus and start writing. Determine how frequently you will write and put it on your calendar. The more frequently you publish to your blog, the better. Each new post is a new page on the web for search engines to catalogue and each is a new search result. Search engines want fresh results and each time you post that exactly what you are providing.

Don't know what to write about? See Resources for a blog I wrote called 15 Easy Blog Post Topics for some great ideas. It's normal to not know what to write about and everyone wants to know. The list of subjects in that blog will help you get started and plan your approach. Blogs don't need to be long! Actually, if they are short, they are more likely to be read. Shoot for 500 words or more.

Use additional social media tools

I know this is surprising but when I started using social media many many years ago, my goal wasn't to dominate the search results associated with my name. Back then social media wasn't even a factor in search. The beauty of using tools such as Facebook, Twitter, LinkedIn, Slideshare, About.me, YouTube, Google+, Flickr, Pinterest, Yelp, Quora, Foursquare etc. is that they provide the fresh and relevant search results search engines salivate over. Your clients or those that are searching for you spend much of their time on social media so it makes sense to be found there as well.

Set up a schedule for monitoring your name or brand

I do this monthly when I close my books. I set up a simple Excel spreadsheet to track the accounts, log in information, and any results I find or follow-up I want to do. If you are an individual you can do this on the first of the month, quarterly, or simply when you think of it. I do recommend setting up some kind of a reminder to make sure you do this fairly regularly. I personally do this at the same time as my metric report.

Set up Google Alerts (http://www.google.com/alerts)

- Create automated alerts to notify an email account when information is posted online (as it happens, once per day, or weekly).
- Base alerts around keywords, brand names, organization names, web addresses, events, and more.
- Remove those that return too many results and edit to get a better return.

Set up saved searches on a variety of other tools

- Hootsuite - Twitter, Facebook, LinkedIn, WordPress, Foursquare and Google+
- Klout – measure of online influence
- Social Mention – sentiment

While you can't make negative search results associated with your name completely disappear, you can displace those results with what you'd like the world to know. Though we've never had complete control over what is said about us, we DO have some control over how we are viewed on the web. Stop sitting back in your chair. Sit forward, put your hands on the keyboard, and craft the message you want them to see.

Read more of my blogs on reputation management: http://bit.ly/krcrepmng

CHAPTER 8

How to Identify the Right Team Member for Social Media Management

S kills for the job - There is an assumption that only "young people" will be able to handle social media. This is **not true**. The most important element I see in successful social media management is someone that is curious and willing to try new things. Peter W. Schutz, the former CEO of Porsche, once said:

"Hire character, train skill."[19]

I couldn't agree more. I'm a trainer. It's rare that I can't teach someone a new skill but I **can't** train employees to be good people.

Most likely you won't need a person full time to handle your social media. The tasks can be added onto a marketing person's list of jobs if they aren't already stretched too thin. This might be a junior marketing person or an administrative person. Part of my job (and all the tips I've given you) will help keep it to a minimum and my goal is to have those I work with to be using it for less than an hour a day. It doesn't have to take all day to manage social media, not unless your brand is so diverse that it requires that kind of time or you have the budget to spend.

The person that you will hire or train will have some, if not all, of the following qualities or skills:

- Passion for, or willingness to learn to use, new technology
- Firm grasp on what services the program offers and what other programs or departments are a good fit as content partners
- Good verbal and written communication skills
- Encouraged by change rather than overwhelmed by it
- A minimum of a half hour to devote to social media management per day
- Flexibility
- Curiosity

Social media is less about the technology and more about the ability and desire to communicate.

Training

Hands-on training is the best way to learn new technology. Self-education is another route but it can take longer and yield poor results. As of the writing of this book, there are only three accredited universities in the U.S. that offer a certification or any kind of degree in social media. I happen to be a co-instructor in one of those courses at Sonoma State University in Rohnert Park, CA.

You will find it very near impossible to hire a person who is within your budget that specializes in social media since the skills are too much in demand right now. It's best to take someone that already works for you and train them by hiring a professional or hiring a part time marketer or assistant that can be trained.

Outsourcing the Job

> *"Forget this! I should just hire an outside firm*
> *to do all this for me!"*

I bet this is what you are thinking right now. Talking to all the people that I've met who have used outside firms, I don't believe you'll get what you want by hiring this way.

A clarification: hiring people to create content (bloggers, technical writers, photographers, etc.) is a good idea and raises the professional level of your output but I'm talking about turning over 100% of your social media management to an outside firm. Let me give you my reasons why I advise against this.

- An outside company is going to charge an arm and a leg to create content for you and post on your behalf. The odds are very good that you aren't going to get a good return on your investment (see Chapter 6 to learn how to calculate this).
- They won't know your industry, its terms, concerns, language, and culture they way you or a team member would. They won't be able to express the appropriate passion, interest, or vernacular to ring authentic. Yes, there are firms that create social media content just for dentists, as an example, but all their dentist clients get the

exact same content so it's actually a waste of money (in my opinion). Even managing accounts for dozens of the same client doesn't give them first hand knowledge of the industry since they aren't actually dentists.

- They commonly will do only the bare minimum to maintain the contract (they have a tremendous amount of work and clients) and your accounts will feel like it – bare minimum.

- They often use filler content that isn't related to your industry or isn't appropriate. This can damage your brand and search engine rankings because your social media channels will be filled with useless posts that are off topic.

- They have access to all of your accounts and login information. Should your relationship sour, they hold the keys to your online reputation. I'm not saying they will be unprofessional but you are taking that risk of relinquishing control of your voice and your brand to someone else. Frequently they set up new accounts for their clients and when a contract ends, the client doesn't know what accounts they own, how to get access, how to use the tools, or how to fix them when the management company does something wrong or ill-advised.

Yes, I'm a social media trainer so I have a vested interest in training you and your staff. But I truly believe you having ownership over your channels, message, and content is the plan for your best benefit. I have heard too many horror stories to trust this process. I want you to be empowered rather than beholden to someone else. But that's just how I roll. You will make your own decision in the matter.

In Closing

I was the lead trainer for a California Colleges initiative some years back and my coordinator, Steve Wright, said something to me that resonates to this day.

> *"40 years ago, when you started a business, you got a location and a phone line and you were in business. No one ever asked the ROI of a phone."*

My advice is to think about your goals, plan ahead, research, use critical thinking, don't obsess over your numbers, and be patient. Let your passion for what you do show through when you create. It makes a big difference. And finally, you CAN do this.

I want to thank you for picking up this book. It follows the process I take my clients and students through to set up a social media campaign for success. I want you to have the information you need, the resources to help you, the tools to make smart decisions, and a system designed for simplicity.

Please check out the following sections for resources and a bibliography for further reading.

I appreciate your time and let me know if I can be of help to you!

My very best to you and yours,
Kerry Rego @kregobiz

kerry@kerryregoconsulting.com

Sales

You may order more copies of this book and get discounts for bulk orders by emailing book@kerryregoconsulting.com.

Speaking

You can invite me to speak in support of this book's subject matter by emailing speaker@kerryregoconsulting.com,

Best Practices

- Confirm you are logged into the correct account before posting.
- Post any public comment policy in a conspicuous location on any social media account or provide the link on a regular basis so that the audience may have the opportunity to read it.
- Review company/organization social media policy before posting.
- Social media is designed to be humans communicating with each other. Try to keep a conversational and natural tone in order to attract viewers and engagement.
- Social media accounts are supplemental to your main points of conversion. The accounts should drive traffic to your owned digital properties.
- Posts should have a call-to-action. These CTAs should support your goals for your campaign. Encourage them to sign up for the mailing list, visit the website, call into the office, sign up for an event – actions that you will count as conversions. This is a marker for success.
- Anytime you receive a message from another account or person that is simply a link or the majority of the message is a link, DON'T click on it. This is often virus-laden and will install malware or has some other

malicious intent. Check with the person (if you have a previous relationship) outside of the suspected message to see if they in fact sent it to you.

- If someone ever asks you if you've sent them a message that you don't recognize, immediately change the account password and check to see what messages have come from that account that may not have been authorized.

- Learn what users on each channel dislike such as direct messages on Twitter, writing in all capital letters, or using abbreviations incorrectly. Knowing what NOT to say is just as important as learning the right things to say.

- Make sure you identify the leaders in your industry and attempt to make authentic conversation with them. You will learn and become better connected in your field.

- Sharing is part of the fabric of social media. Cite your sources whenever possible by identifying the source, tagging, linking or whatever is appropriate. Give credit where credit is due. This also helps to build relationships with those whose work you share.

Resources

Templates

Visit http://bit.ly/SMSKtemplates for all downloads

- Social media strategy
- Editorial calendar monthly template
- Editorial calendar individual template
- Digital assets template
- 15 Easy Content Topics
- Empty social media house
- Metric report template, simple
- Metric report template, advanced
- Timely Task List
- Crisis plan template
- Photo/media recording release
- OR buy the workbook (includes all templates) http://www.kerryregoconsulting.com/store

Statistics and information

- Demographic data, Pew Research Center. http://www.pewinternet.org/
- Website traffic, comScore. http://www.comscore.com
- Website traffic and rankings, Alexa. http://www.alexa.com

- "CAN-SPAM Act: A Compliance Guide for Business," Bureau of Consumer Protection http://bit.ly/CANSPAMAct
- "Blog Readership Demographics," Royal Pingdom. http://bit.ly/BlogRdr
- Maeve Duggan and Aaron Smith. "Social Media Update 2013", Pew Research Center. http://bit.ly/PewSocM2013
- "Why Use Visual Content in Social Media Marketing?", B2B Infographics. http://infographicb2b.com/2013/07/03/why-use-visual-content-in-social-media-marketing-infographic/
- "10 Surprising Social Media Stats To Make You Rethink Your Strategy," Buffer. http://blog.bufferapp.com/10-surprising-social-media-statistics-that-will-make-you-rethink-your-strategy

Media

Free and Royalty-free Images

- Wikimedia Commons & Public Doman Royalty Free Photography. http://commons.wikimedia.org/wiki/Commons:Free_media_resources/Photography
- Flickr, The Commons.

http://www.flickr.com/commons/
- Pexels. http://www.pexels.com

Stock Images Sites, Cost

- iStockPhoto. http://www.istockphoto.com
- Getty Images. http://www.gettyimages.com

Photo editing resources. http://bit.ly/imagetools

Royalty-free music. http://freeplaymusic.com

Microsoft photo release template.
http://office.microsoft.com/en-us/templates/photographer-s-image-use-release-form-TC103988279.aspx

Marketing

Marketing, content, and specific tool information.

- Mashable. http://mashable.com
- TechCrunch. http://techcrunch.com
- Hubspot. http://www.hubspot.com

Return on Investment (ROI)

- "A Simple Way to Calculate Social Media Return on Investment," Social Media Examiner. http://www.socialmediaexaminer.com/a-simple-way-to-calculate-social-media-return-on-investment/
- "Social Media ROI: 14 Formulas to Measure Social Media Benefits," Search Engine Watch. http://searchenginewatch.com/article/2249515/Social-Media-ROI-14-Formulas-to-Measure-Social-Media-Benefits
- "The Beginner's Guide to Calculating Social Media ROI," Brandwatch. http://www.brandwatch.com/2014/03/the-beginners-guide-to-calculating-social-media-roi/
- "Analytics Tip #17: Calculate Customer Lifetime Value (LTV)," Occam's Razon by Avinash Kaushik. http://www.kaushik.net/avinash/analytics-tip-calculate-ltv-customer-lifetime-value/

Online Training & Employee Policies

- YouTube – check timestamp on video
- GCF LearnFree http://www.gcflearnfree.org
- Lynda (cost) - http://www.lynda.com
- Social Media Governance-
  http://socialmediagovernance.com/policies

Bibliography

1. ".com Disclosures," Federal Trade Commission. http://www.ftc.gov/sites/default/files/attachments/press-releases/ftc-staff-revises-online-advertising-disclosure-guidelines/130312dotcomdisclosures.pdf

2. "SMART Criteria," Wikipedia, last modified September 17, 2014, accessed September 18, 2014. http://en.wikipedia.org/wiki/SMART_criteria

3. "Understanding Keywords," Hubspot. http://knowledge.hubspot.com/keyword-user-guide/understanding-keywords

4. "CAN-SPAM Act: A Compliance Guide for Business," Bureau of Consumer Protection, accessed September 1, 2014. http://bit.ly/CANSPAMAct

5. "Why Use Visual Content in Social Media Marketing?", B2B Infographics. http://infographicb2b.com/2013/07/03/why-use-visual-content-in-social-media-marketing-infographic/

6. Demographic data, Facebook. http://newsroom.fb.com/company-info/

7. "Twitter's Users Are in Asia, but Its Revenue is in the U.S.," BusinessWeek. http://www.businessweek.com/articles/2014-05-27/twitters-users-are-in-asia-but-its-revenue-is-in-the-u-dot-s-dot

8. Demographic and user information, YouTube. https://www.YouTube.com/yt/press/statistics.html

9. "YouTube: The 2nd Largest Search Engine," Visual.ly. http://visual.ly/YouTube-2nd-largest-search-engine

10. "Search Engine Optimization," Wikipedia, last modified September 11, 2014, accessed September 17, 2014. http://en.wikipedia.org/wiki/Search_engine_optimization

11. "Dark Social: We Have the Whole History of the Web Wrong," The Atlantic.

http://www.theatlantic.com/technology/archive/2012/10/dark
-social-we-have-the-whole-history-of-the-web-
wrong/263523/

12. "PPM Coalition Calls Arbitron Data Presentation Inaccurate,"
HispanicAd.com. http://hispanicad.com/blog/news-
article/had/research/ppm-coalition-calls-arbitron-data-
presentation-inaccurate

13. "Arbitron's Chief Resigns After a False Statement," The New
York Times.
http://www.nytimes.com/2010/01/13/business/media/13arbit
ron.html?_r=0

14. "Nielsen Method for TV Ratings Missing Minorities, Young
People," Poynter. http://www.poynter.org/latest-
news/measuring-audience/225876/nielsen-method-for-tv-
ratings-missing-minorities-young-people/

15. "Nielsen Admits Undercounting Web Traffic," Advertising Age.
http://adage.com/article/digital/digital-nielsen-admits-
undercounting-web-traffic/146899/

16. " Be Transparent: Why Deleting Negative Posts is a Bad Idea,"
Kerry Rego Consulting.
http://kerryregoconsulting.com/2012/06/22/be-transparent-
why-deleting-negative-posts-is-a-bad-idea/

17. Chris Brogan and Julien Smith. Trust Agents.
http://www.amazon.com/Trust-Agents-Influence-Improve-
Reputation/dp/0470635495

18. "How To Get Started In Reputation Management," Kerry Rego
Consulting http://kerryregoconsulting.com/2012/03/09/how-
to-get-started-in-reputation-management/

19. "The Big Apple: 'Hire Character. Train Skill,'" Barry Popik.
http://www.barrypopik.com/index.php/new_york_city/entry/
hire_character_train_skill

ABOUT THE AUTHOR

Kerry Rego is a social media trainer, technology consultant, and keynote speaker that works with individuals, businesses, government, and non-profits. She educates, implements, and trains people of all ages on new media.

Kerry is the County of Sonoma social media staff trainer, Sonoma State University Extended Education instructor, Santa Rosa Junior College Community Education instructor, Windsor High School guest instructor, Napa-Sonoma Small Business Development Center instructor, and a North Bay Business Journal columnist. She is a recipient of the North Bay Business Journal's 2011 "Forty Under 40 Award" and the author of *What You Don't Know About Social Media CAN Hurt You: Take Control of Your Online Reputation.*

Kerry is based out of Santa Rosa, CA in beautiful Sonoma County, an hour north of San Francisco. She was born there and makes it her home with her husband and daughter.